Acute effects on health of smog episodes

WHO Library Cataloguing in Publication Data

Acute effects on health of smog episodes : report on
a WHO meeting, 's Hertogenbosch, Netherlands,
30 October – 2 November 1990

(WHO regional publications. European series ; No. 43)

1.Environmental health 2.Air pollution – adverse effects
3.Smog – adverse effects 4.Risk factors 5.Europe
I.Series

ISBN 92 890 1306 0 (NLM Classification: WA 754)
ISSN 0378-2255

World Health Organization
Regional Office for Europe
Copenhagen

Acute effects on health of smog episodes

Report on a WHO meeting

's Hertogenbosch, Netherlands
30 October – 2 November 1990

WHO Regional Publications, European Series, No. 43

ICP/CEH 098
Text editing by Frank Theakston

ISBN 92 890 1306 0
ISSN 0378-2255

PRINTED IN ENGLAND

CONTENTS

Introduction

A Working Group on Acute Health Consequences of Winter-type and Summer-type Smog Exposures was convened from 30 October to 2 November 1990 in 's Hertogenbosch, the Netherlands. The meeting was held in collaboration with the National Institute of Public Health and Environmental Hygiene in Bilthoven, and with the financial assistance of the Dutch Ministry of Housing, Physical Planning and Environment. It was attended by eighteen temporary advisers from eight countries. Dr M. Lippmann was elected Chairman and Dr B. Brunekreef Rapporteur. Dr M.J. Suess was Scientific Secretary to the meeting.

The Working Group was convened because during episodes of smog air quality guidelines for air pollutants of major importance can be exceeded to the extent that acute adverse effects on health may occur. Such episodes happen during stagnant weather conditions both in summer and in winter, though the pollutants of primary concern during winter and summer episodes usually differ. Increasing concern by the general public about the possible health consequences of smog episodes has led to demands for appropriate action by the authorities. In response, various "smog alert systems" and abatement strategies have emerged. These, however, differ considerably from country to country, even though episodes of smog may cover several countries at the same time, with comparable levels of pollution.

The Working Group restricted itself to evaluating the short-term effects on health of acute and episodic exposures.

This does not mean to imply, however, that episodic exposures may not have long-term consequences for human health.

Differences in the evaluation of the health effects in smog situations reduce the credibility of control policies that are based on the assessment of health effects. There is thus a need to quantify the exposure of the population to air pollution during smog episodes and the risks to health associated with it. Also, there is a need to advise the population on suitable behaviour during episodes of smog.

Discussion concentrated on assessing the risks to health of exposure to elevated concentrations of air pollutants during episodes of smog in winter and summer. In winter, episodes may occur during stagnant weather conditions when pollutants generated by the burning of fossil fuel accumulate in the atmosphere. The pollutants of primary interest are sulfur dioxide and suspended particulate matter, though these merely serve as indicators of a much more complex mixture of pollutants. In summer, episodes may occur during warm, sunny weather; photochemical reactions of nitrogen oxides and hydrocarbons in the atmosphere lead to the formation of ozone and other harmful substances.

For winter-type smog, information on expected acute health effects was derived primarily from epidemiological studies. The principal indicator components of winter-type smog – sulfur dioxide, suspended particulate matter and sulfuric acid – have not been shown experimentally to cause significant effects on human health by themselves, in the concentrations typically encountered in recent episodes. For summer-type smog, ozone is considered to be the major indicator component. Experimental evidence shows that ozone induces measurable, reproducible and statistically significant changes in lung function indices at exposure levels commonly encountered in such episodes. Information on expected acute health effects was therefore derived from human experiments, and also from epidemiological studies of acute responses to the mixture of pollutants present in summer-type smog.

For assessing the risk to health of exposure to these smog mixtures, the expected effects on health were classified as "mild", "moderate" or "severe". When effects are expected to be moderate, some public advice on exposure or dose reduction for sensitive individuals could be considered. When effects are expected to be severe, additional measures can be recommended on a voluntary basis, and emergency short-term measures such as closing schools or limiting road traffic can be considered.

It was generally felt that reducing the baseline levels of pollution is the preferred and most efficient way of reducing exposure to air pollutants during episodes. Reducing the emission of primary pollutants during episodes will usually not lead to a proportional reduction in population exposures, as the contribution of episodes to long-term total exposure is relatively small. Moreover, secondary pollutants are thought to be the most harmful to health.

Winter-type and summer-type smog

Winter-type Smog

For the purpose of this report, winter-type smog chiefly means pollution from the combustion of sulfur-containing fossil fuel for heating and/or energy generation. Under stagnant weather conditions, sulfur dioxide and suspended particulate matter may accumulate in the atmosphere and react to form sulfuric acid. When the weather is cold energy demand increases, with increased emissions as a result. When there is snow cover, dry deposition of pollutants is greatly reduced so that less pollution is scavenged from the air. Mixing height may also be reduced. These circumstances occasionally combine, leading in the past to very high concentrations of pollutants in large conurbations such as London, but also more recently in areas of eastern Europe. Relatively high concentrations have been observed in western Europe as well, as recently as the winters of 1985 and 1987, as a result of long-range transport over several hundred kilometres. The 1985 episode has been studied in detail *(1)*; it lasted from 14 to 22 January, and was caused by a high pressure system over Scandinavia and eastern Europe. Winds were from the east and south-east in western Europe, extensive snow covered much of central and western Europe, and cold air masses were trapped in the lower boundary layer by warmer air brought by southerly winds. Trajectory analyses showed that much of the pollution measured in western Germany and the Netherlands came from large source areas in eastern Europe. It was estimated that at the eastern border of the Netherlands,

55% of the sulfur dioxide came from eastern Europe and a further 25% from western Germany. The highest 24-hour average sulfur dioxide concentrations observed at Kassel in Germany were just over 1000 μg/m^3 *(2)*. In the same episode, 24-hour average concentrations of more than 2000 μg/m^3 were observed in Leipzig. In the Netherlands, sulfur dioxide concentrations were generally below 500 μg/m^3. Information on suspended particulate matter concentrations is scarce. Data indicate that the highest 24-hour average concentrations observed in Germany were of the order of 500 – 600 μg/m^3 during this episode *(2,3)*.

As winter-type smog episodes usually coincide with cold weather, people will tend to remain indoors much more than during summer-type episodes. Staying indoors offers protection from outdoor pollutants such as sulfur dioxide that react with indoor surfaces *(4)*. However, a study on indoor–outdoor relationships conducted in the Netherlands during the episode of 1985 indicated that respirable particulate matter penetrated well into homes, suggesting that indoor exposure contributed considerably to total exposure to particulate pollution during that episode *(5)*.

Unfortunately, specific episode evaluations from eastern Europe have not been published. Data from Poland presented at the Workshop indicated that in 1987 (when another winter episode took place), the maximum 24-hour average sulfur dioxide concentration observed in the heavily industrialized province of Katowice was about 1300 μg/m^3. In the same province, the maximum daily average for suspended particulate matter was about 1200 μg/m^3. During the 1952 London episode, maximum levels of sulfur dioxide and particulate matter (measured as smoke) both reached about 5000 μg/m^3.

Few data exist on the concentrations of other air pollution components of winter-type smog. Aerosol acidity has been measured in London since 1959, and in those early years, maximum 1-hour average concentrations of sulfuric acid of up to 680 μg/m^3 have been observed *(6)*. In recent episodes, 24-hour average sulfate levels have peaked at just over 100 μg/m^3 in the Netherlands *(7)* and, due to neutralization, sulfuric acid levels have certainly been much lower.

Depending on local circumstances with respect to emission characteristics, topography and climate, the pollution mix may vary from place to place in winter-type episodes. No single indicator pollutant (or combination of indicator pollutants) can therefore be used to predict exactly the health consequences of exposure to winter-type smog under different circumstances.

Summer-type Smog

For the purpose of this report, summer-type smog refers primarily to photochemical pollution arising from atmospheric reactions of hydrocarbons and nitrogen oxides, stimulated by intense sunlight. Ozone is considered the most biologically active pollutant within this mixture. However, when locations at different latitudes with different solar ultraviolet intensity and different source contributions are compared, the toxicity of the photochemical smog at the same ozone concentrations is likely to be substantially different. Also, other components than ozone are known to be responsible for eye irritation and other annoyance effects *(8)*; although the ozone concentration is correlated with these effects, its usefulness as a pollution variable for these effects is questionable. Owing to the large variety of sources and atmospheric reactions, air pollution in large cities is extremely complex *(9)*. Air pollution exposures are poorly characterized in these situations, and the risks to health of exposure to complex mixtures of air pollutants are still poorly understood.

Ozone exposures in summer-type episodes usually follow a typical diurnal pattern, with peak concentrations in the afternoon and low levels during the night and early morning. This is related to the high reactivity of ozone, which reduces in concentration at ground level during the evening and night, and to scavenging by nitric oxide. However, above the mixing layer, high ozone concentrations may prevail throughout the night. When mixing increases during the morning, ground level concentrations of ozone can increase as a result, so that peak ozone concentrations may increase from day to day. Owing to scavenging by other pollutants, ozone concentrations in city centres are often lower than in areas downwind of large sources or source areas. Ozone concentrations generally increase during warm, sunny weather, when people tend to spend more time outdoors. High exposures may thus result, especially when people are physically active. As ozone is extremely reactive, concentrations indoors are usually much lower than those outdoors, although they have been found to vary widely *(10)*. The ratio of personal ozone exposure to the average outdoor concentration has been estimated to fall within a range of 0.6 – 0.8 for most population groups *(11)*.

Even though traffic emissions of nitrogen oxides and hydrocarbons constitute a major proportion of ozone precursors in North America and western Europe, the immediate reduction of emissions through traffic bans and the like are not expected to result in large falls in peak concentrations. For example, it has been estimated that peak concentrations of ozone in the Netherlands would be reduced by 4% on average if all traffic in that country stopped *(12)*. Locally, larger or smaller reductions or even increases may occur. If all European traffic were stopped, it is estimated that peak concentration would fall by 26%. These calculations illustrate that local, short-term measures to reduce emissions during summer-type smog episodes are relatively ineffective in reducing ozone exposure, and that prevention of future episodes by measures to reduce baseline precursor emissions is the more sensible and effective course of action.

Effects on health of winter-type smog

Epidemiological studies have shown that exposure to winter-type smog is associated with a range of effects on human health. Observed effects have included temporary changes in pulmonary function, an increase in morbidity among chronic bronchitics, an increase in hospital admissions due to respiratory and cardiovascular conditions and, depending on the severity and nature of the exposure, increases in mortality.

Ever since excess numbers of deaths were found to be associated with episodic air pollution exposures in such places as London and the Meuse Valley in Belgium, the relationship between mortality and air pollution has been the subject of much study. In particular, the large number of some 4000 deaths associated with the London smog episode of December 1952 provided much impetus to emission reduction measures in the United Kingdom and elsewhere. Concentrations of sulfur dioxide, black smoke and presumably sulfuric acid were very high in the 1952 London episode; such high concentrations have not been observed in recent episodes.

Several authors have tried to determine a threshold of 24-hour average sulfur dioxide and/or particulate matter concentrations below which no significant effect on mortality occurs. Reviewing these attempts, WHO recently concluded *(8)* that the lowest observed effect level for increased mortality due to winter-type smog exposure can be put at 24-hour average concentrations of 500 $\mu g/m^3$ sulfur dioxide combined with 500 $\mu g/m^3$ black smoke. Realizing the difficulty of trying to define such a threshold precisely, it was stated that "This does not preclude the possibility that mortality effects occur below these concentrations".

Recent studies continue to suggest that at relatively low levels of sulfur dioxide, associations between mortality and air pollution can be observed *(13 – 15)*. An epidemiological study from Athens *(13)* has suggested that the threshold for effects of the local pollution mix (as characterized by sulfur dioxide) on mortality lies below 150 μg/m^3. A further analysis of these data compared mortality on days with sulfur dioxide concentrations over 150 μg/m^3 with that on days with concentrations below 150 μg/m^3, matched for temperature, three-year period, season, day of the week and holidays *(14)*. Respiratory mortality on high exposure days was significantly greater than on comparison days. High exposure days included days with sulfur dioxide concentrations of up to 940 μg/m^3 and black smoke concentrations of up to 790 μg/m^3.

A study from two cities in southern France documented associations between respiratory mortality and urban air pollution as characterized by sulfur dioxide concentrations *(15)*. Monthly mean concentrations were up to 200 μg/m^3 in the investigated winters, but 24-hour average concentrations were said not to have been higher than 500 μg/m^3. Long-term average suspended particulate matter concentrations were higher than sulfur dioxide levels, but it was not reported how high the maximum 24-hour concentrations were in the winter. A study from Germany has associated a small increase in mortality during the 1985 episode with maximum 24-hour average sulfur dioxide concentrations over 800 μg/m^3 and suspended particulate matter concentrations of over 600 μg/m^3 *(3)*. Whereas the German data seem to fall in line with the earlier WHO assessment *(8)*, the studies from southern Europe suggest that the pollution mix may be somewhat more harmful at a given sulfur dioxide concentration than that associated with the mortality noted in studies from north-western Europe.

Exposure to winter-type smog has been associated with increased morbidity. As an example, the United Kingdom panel studies among chronic bronchitics can be cited *(16)*, in which exacerbation of the patients' condition was found to be associated with 24-hour average black smoke concentrations of over 250 μg/m³ and sulfur dioxide concentrations of over 500 μg/m³. Increases in hospital admissions for respiratory conditions were observed in the German study of the 1985 episode *(3)*. In the winter of 1984/1985, which included the January 1985 episode, a Berlin study *(17)* showed hospital admissions for croup syndrome (acute stenosing laryngo-tracheitis) to be associated with sulfur dioxide concentration on the day before admission. In general, air pollution on the days before admission was higher than during the days directly after admission (i.e., the frequency of days with 24-hour means for sulfur dioxide exceeding 300 and 400 μg/m³ respectively). Little association was found between treatment for respiratory conditions in casualty departments and air pollution in the industrial town of Steubenville, Ohio, USA *(18)*. Maximum 24-hour average sulfur dioxide concentrations were up to 370 μg/m³, and maximum total suspended particulate matter concentrations were up to 700 μg/m³ in this study.

Transient changes in pulmonary function have been associated with exposure to episodic winter-type smog exposures in both children and adults. In an early study in the Netherlands *(19)* adults were found to have higher lung function levels in 1972, three years after an initial measurement. This was associated with exposure to relatively high concentrations of sulfur dioxide and black smoke during the initial measurements of lung function: 24-hour average sulfur dioxide levels were from 200 to 300 μg/m³, and black smoke levels were from 100 to 150 μg/m³. In November 1975, an air pollution episode occurred in Pittsburgh, USA *(20)*, the maximum 24-hour average concentration of particulate matter being 770 μg/m³. Lung function was measured in over 200 schoolchildren each day for one week starting at the end of the episode. In most children, FVC and $FEV_{0.75}$ increased over this period, suggesting a temporary decrease associated with the episode. In a study conducted in Steubenville, Ohio *(21)*, temporary decreases of FVC and $FEV_{0.75}$ were associated with elevated levels of sulfur dioxide and particulate matter. Averaged over 24 hours, maximum concentrations were 280 μg/m³ sulfur dioxide and 420 μg/m³ particulate matter in one episode, and 460 μg/m³ and 270 μg/m³, respectively, in another. The reductions in lung function were not more than about 5% of the group mean. Further analysis of the data from this study showed that there was no evidence for heterogeneity of response in the studied population of schoolchildren *(22)*. This suggests that in this population, there were no children with a markedly stronger response to air pollution exposure than the group mean. In the 1985 episode in central and western Europe, a small temporary decline in FVC, FEV_1 and some expiratory flow measures was observed in a group of schoolchildren studied in the Netherlands *(23)*. Concentrations of sulfur dioxide and particulate matter (24-hour averages) were in the range 200 – 250 μg/m^3 for a number of days. Respirable suspended particles with a cut-off at a mass median diameter of 3.5 mm were also measured, and their concentration was found to be almost equal to that of the particulate matter. This suggested that almost all the dust in the air was in the fine particulate state, which also helps to explain the relatively efficient penetration of these particles into homes *(5)*. The decline in lung function was not more than about 5% on a group mean basis, and follow-up measurements suggested that lung function had returned to baseline levels about three weeks after the episode.

During the 1987 European episode, lung function was measured in a group of patients with moderate airway obstruction living in the Federal Republic of Germany *(24)*. The maximum 24-hour average sulfur dioxide concentration observed in the study area was 540 μg/m^3. The FVC and FEV_1 were lower than baseline levels by 5% and 7%, respectively. In the same episode, schoolchildren were studied in the Netherlands *(25)*. Again, a small transient decline was observed in FVC, FEV_1 and measures of expiratory flow. The highest 24-hour average sulfur dioxide concentration in the study area was about 300 μg/m^3. Black smoke concentrations did not exceed 100 μg/m^3, but the maximum 24-hour average particulate matter concentrations were almost 300 μg/m^3. On one day, sulfate concentrations in the centre of the country reached 130 μg/m^3 as a 24-hour average.

There is little experimental evidence implicating individual pollutants as causal agents for the health effects observed in epidemiological studies of winter-type smog exposures.

Asthmatics are known to be more sensitive to sulfur dioxide than healthy people, but below 1000 μg/m³ there is little experimental evidence to suggest significant bronchoconstriction among exercising asthmatics *(8)*. Even though these experiments were using short-term (10-minute) exposures, significant lung function decrements have been observed among healthy people associated with 24-hour average exposures as low as 200 – 250 μg/m³ during smog episodes, and short-term peak concentrations have not reached 1000 μg/m³ in these episodes. Asthmatics are also more sensitive to inhalation of sulfuric acid: chamber studies have suggested that among exercising asthmatics, lung function decrements can be induced by short-term exposures to 100 μg/m³ of sulfuric acid. It has been suggested that cumulative exposures to acidic aerosols may be more relevant than the actual concentrations during short periods of time, so that prolonged exposure to tens of micrograms of sulfuric acid per cubic metre during smog episodes could result in dosages to the respiratory tract higher than those that have given significant effects in chamber studies *(26)*. This would not, however, directly explain the effects seen in healthy children.

For the time being, epidemiological studies continue to provide the most crucial evidence for health effects associated with exposure to winter-type smog. Transient effects on lung function have been observed at 24-hour average concentrations of about 200 μg/m³ for sulfur dioxide and particulate matter. At higher levels, the condition of chronic bronchitics has been shown to worsen, and at levels over 500 μg/m³ for both sulfur dioxide and particulate matter, an increase in mortality among susceptible population groups can be expected. The observations have been made largely in northwestern Europe and in the eastern United States. One also has to bear in mind that in other circumstances, the toxicity of the pollution mix may be different, so that effects on health will occur at levels of sulfur dioxide and suspended particulate matter in air other than those given in this report.

Effects on health of summer-type smog

Ozone is considered biologically to be the most active component of photochemical or summer-type smog. Not all effects on health associated with exposure to summer-type smog can be ascribed to ozone, or to ozone alone, however. This is true especially for annoyance effects, for example eye irritation, ascribed to non-ozone irritant components such as organic nitrates and aldehydes *(8)*. However, no reliable dose – response information exists other than the observation that these effects begin to occur when ozone levels of approximately 200 μg/m^3 are exceeded. As these annoyance effects are not due to ozone, however, they may occur at much lower levels of ozone in situations where ozone is scavenged from the air by other pollutants, as on busy streets. Within large conurbations with complex sources of air pollution, scavenging of ozone may lead to relatively low ozone concentrations; under these circumstances, it will not be a reliable indicator for the health risks associated with exposure to summer-type smogs as they occur in such areas.

For the purpose of this report, the discussion on health effects of exposure to summer-type smog will be restricted largely to a discussion of the effects on health of ozone, alone or as a major component of the pollution mix. In the absence of sufficient data, no quantitative evaluation is possible of health risks associated with exposure to summer-type smogs that are not well characterized by the concentration of ozone in the air.

Comparison of results from experimental and epidemiological studies suggests that ozone is the major cause of the health effects of summer-type smogs as they occur in Canada and the United States *(27)*. There have been few epidemiological studies of the health effects of photochemical pollution in Europe.

Early experimental studies have emphasized short-term exposures (1 – 2 hours) following observations of relatively sharp daily peak concentrations of ozone in the atmosphere. In recent years, we have come to realize that in heavily populated areas such as the eastern United States and western Europe, maximum 8-hour average concentrations are often as high as 90% of the one-hour peak concentrations *(28)*. As a result, attention has shifted to evaluation of multi-hour or even multi-day exposures.

Effects of ozone on lung function, bronchial reactivity, exercise performance and respiratory and other symptoms have been documented in experimental studies on humans. Effects of summer-type photochemical smog on lung function, symptoms and hospital admissions have been found in epidemiological studies. Associations with mortality have not been unequivocally shown. Some typical examples of studies in this field are discussed below.

Several investigators have evaluated the effects of exposure to photochemical smog in the Los Angeles basin. Human volunteers were exposed during heavy exercise for one hour to clean air, to ozone alone and to Los Angeles smog characterized by an ozone concentration of about 300 $\mu g/m^3$ *(29)*. After smog or ozone exposure, FVC and FEV_1 were lower than in those exposed to clean air. The magnitude of the response was in the order of – 0.5 to – 1.0 ml per $\mu g/m^3$ ozone, either alone or as a component of the photochemical smog mixture. Higgins et al. *(30)* studied children visiting a summer camp in the Los Angeles basin. Maximum one-hour ozone concentrations varied from 40 to 490 $\mu g/m^3$ in the study period, and a significant relationship was found between ozone and lung function. The magnitude of the effect was estimated to be – 0.2 ml per $\mu g/m^3$ for FVC and FEV_1. Studies conducted in the eastern United States have generally suggested stronger responses to summer-type smog at a given level of ozone. In a study among exercising adults who were exposed for about 30 minutes to air pollution characterized by ozone concentrations of 40 – 250 μg/m^3, FVC and FEV_1 responses were estimated to range from – 0.7 to – 1.5 ml per μg/m^3 ozone *(31)*. In a study among children who were visiting a summer camp, one-hour maximum ozone concentrations ranged from 40 to 225 μg/m^3 *(32)*. The estimated FVC and FEV_1 responses were – 0.5 and – 0.7 ml per μg/m^3 respectively. It has been suggested that the observed stronger responses in the eastern United States are a result of exposure to other air pollution components such as sulfuric acid *(27)*.

The lung function response to ozone has been shown to increase with exposure time. A 6.6-hour exposure of exercising adults to 240 μg/m^3 ozone produced an effect on FVC of – 1.9 ml per μg/m^3, and an effect on FEV_1 of – 2.3 ml per μg/m^3 *(33)*. In experiments of this length, significant effects on lung function have been demonstrated at ozone concentrations as low as 160 μg/m^3 *(34)*.

Physical exercise is another determinant of the magnitude of the lung function response to a given ozone concentration, as it increases the dose delivered to the airways and the lung. Most studies on ozone have used some kind of exercise protocol, so that results cannot be directly extrapolated to people at rest. For example, in a study among healthy and asthmatic adolescents at rest, exposed to 240 μg/m^3 ozone, no effects on lung function were observed *(35)*.

In most of the experimental and epidemiological studies among adults, lung function responses were shown to be accompanied by increased reports of respiratory and other symptoms by adults, but not by children. Among the frequently reported symptoms are cough, shortness of breath and pain on deep inspiration *(27)*. In a re-analysis of the results of chamber studies, Ostro et al. *(36)* have suggested that at ozone exposures resulting in a 10% decrease in FEV_1, the probability of experiencing a lower respiratory symptom increases by 15%. In a diary study among student nurses living in Los Angeles, increased reports of chest discomfort were found when one-hour maximum ozone levels were over 400 μg/m^3 *(37)*. In another diary study from the Los Angeles area, a relative risk of 1.4 for reporting an asthma attack was found to be associated with an ozone level of 400 μg/m^3 *(38)*.

Hospital admissions for respiratory conditions in southern Ontario were found to be associated with temperature and with increased though relatively low levels of sulfur dioxide and ozone in summer *(39)*. It has been suggested that co-exposure to acid air pollutants played an important role in this situation *(26)*.

Bronchial responsiveness to stimuli such as metacholine has experimentally been shown to be enhanced by exposure to ozone. For example, in the study by Horstman et al. *(34)* bronchial responsiveness to a metacholine challenge was shown progressively to increase with 6.6 hours of exposure to 160, 200 and 240 μg/m^3 ozone. Also, Koenig et al. have shown that after 45 minutes of exposure to 240 μg/m^3 ozone with moderate exercise, adolescent asthmatics were more responsive to a 15-minute exposure to sulfur dioxide at 290 μg/m^3 *(40)*. These studies suggest that people who already have reactive airways are at greater risk of airway narrowing due to irritative or allergic stimuli after exposure to ozone.

It has also been shown experimentally that airway permeability is increased by exposure to ozone, and that inflammatory changes occur in the lung. For example, Koren et al. *(41)* have documented greater numbers of polymorphonuclear leukocytes in bronchoalveolar lavages from exercising volunteers exposed to ozone for 2 hours at 800 μg/m^3. Later experiments showed that a 6.6-hour exposure of exercising volunteers to 200 μg/m^3 also produces this effect *(42)*.

In chamber experiments, people suffering from a variety of respiratory conditions have not been shown to be more responsive to ozone. Some people are consistently more responsive to ozone than others, but it is not yet clear why this is so *(27)*. For the time being, those who take exercise outdoors are considered to be particularly at risk, as they receive higher doses than others.

In summary, experimental evidence shows that ozone is capable of producing several types of effect on human health after exposure of exercising individuals lasting from one to several hours. The ozone concentrations at which these effects have been observed are well within the range regularly observed in summer-type smogs. Several of the effects (lung function responses, symptomatic responses) observed in these experiments have been found in epidemiological studies as well. In some studies, the magnitude of the effect has been comparable to that found in experimental studies at similar estimated ozone doses. This suggests that in those situations, ozone is the main component of the summer-type smog mixture responsible for the type of effects on human health seen in the experiments. In other studies, notably from the eastern United States, the smog mixture has been shown to cause a larger response at a given ozone concentration than one would expect from the experimental results. It has been suggested that co-exposure to other pollutants such as sulfuric acid is responsible for this.

Health risk assessment

In 1987, WHO issued air quality guidelines for a number of substances *(8)*, including sulfur dioxide, suspended particulate matter and ozone. The guidelines were developed "to provide a basis for protecting public health from adverse effects of air pollution". Specifically, the guidelines "either indicate levels combined with exposure times at which no adverse effect is expected concerning noncarcinogenic endpoints, or they provide an estimate of lifetime cancer risk arising from those substances which are proven human carcinogens or carcinogens with at least limited evidence of human carcinogenicity". It is clear that the guidelines are meant to prevent all adverse effects on human health from air pollution exposure. Consequently, they are set at levels that are sometimes considerably exceeded during the typical winter- and summer-type smog exposures that are the subject of this report. For example, the guideline values for combined exposure to sulfur dioxide and particulate matter, averaged over 24 hours, are 125 $\mu g/m^3$ and 120 $\mu g/m^3$, respectively. For black smoke and thoracic particles, values of 125 and 70 $\mu g/m^3$, respectively, are specified. "Thoracic" refers to particles having a 50% cut-off point at 10 mm diameter. The guideline values for ozone were set at 150 – 200 $\mu g/m^3$ for a one-hour average, and at 100 – 120 $\mu g/m^3$ for an eight-hour average. Consequently, adverse effects on human health are possible, but the guidelines were not developed to assess the extent of the health risks associated with these exposures. This calls for a separate gradation of the health effects known or expected to occur due to winter- or summer-type smog exposures at certain concentrations of indicator pollutants.

Gradation of the health effects of air pollution has been the subject of some discussion in the past. The US Environmental Protection Agency has attempted to grade lung function responses to respiratory irritants into different classes of adversity *(43)*. An adaptation of this scheme is reproduced in Table 1.

Table 1. Gradation of acute lung function, symptomatic and other responses to air pollution exposure into different classes of adversity

Response	Gradation		
	Mild	Moderate	Severe/incapacitating
Change in FVC or FEV_1	5 – 10%	10 – 20%	20 – 40%/> 40%
Symptoms	Mild to moderate cough	Mild to moderate cough, pain on deep inspiration, shortness of breath	Repeated/severe cough, moderate to severe pain on deep inspiration and shortness of breath; breathing distress
Limitation of activity	None	Few individuals choose to discontinue activity	Some/many individuals choose to discontinue activity

Source: Lippmann *(27,43)*.

In this report the terms "mild", "moderate" and "severe" are used to grade expected effects on health of exposure to winter-type and summer-type smog.

In 1985, the American Thoracic Society (ATS) published guidelines on what constitutes an adverse respiratory health effect *(44)*. According to these, mortality, morbidity and pathophysiological changes are considered to be "adverse health effects", whereas physiological changes of uncertain significance and pollutant burdens as such are not. The ATS guidelines considered both acute and chronic effects. Listed as "adverse respiratory health effects" were medically significant physiological or pathophysiological changes generally distinguished by, among others, interference with the normal activity of the affected person, episodic respiratory illness and incapacitating illness. The five most important adverse respiratory health effects listed were, in order of decreasing severity:

- increased mortality;
- greater incidence of cancer;
- higher frequency of symptomatic asthmatic attacks;
- greater incidence of lower respiratory tract infections; and
- increased exacerbation of disease in those with chronic cardiopulmonary or other disease that could be reflected in a variety of ways:
    - — reduced ability to cope with daily activities (shortness of breath or increased anginal episodes)
    - — greater frequency and duration of hospitalization
    - — greater frequency of visits to the physician or hospital casualty department
    - — increased pulmonary medication
    - — reduced pulmonary function.

At the time, small transient reductions in lung function not associated with an asthmatic attack were considered to be of minor importance. It was also recognized that selection of a point on a dose–response curve that separates a medically significant from a medically insignificant effect may be difficult. The ATS guidelines were not developed specifically to address the evaluation of health risks associated with episodic winter- and summer-type smog exposures. They are summarized here to provide some background to the assessment made by the Working Group.

To the knowledge of the Working Group, there have been no previous attempts in the published literature to grade all known or expected acute health effects of winter- and/or summer-type smog exposures in classes of increasing severity. To date, the gradation of a number of health effects of ozone by Lippmann *(27,43)* is the closest approximation of what the Working Group set out to achieve.

Any gradation of health effects has an arbitrary element to it. First, scientific knowledge evolves, and new information on the health effects of the exposures under consideration may become available in the future, leading to a further evaluation of the health risks. Second, the interpretation of available knowledge may change as we learn more about, for example, any long-term consequences of seemingly trivial acute changes. Also, our conception of what is an adverse effect on health depends on our conception of health, which can never be completely scientific or value-free. With these considerations in mind, the Working Group discussed the available information on health effects of winter- and summer-type smog exposures, and graded these effects into classes of increasing severity to the best of its experience and ability.

For winter-type smog, the expected health effects are graded in Table 2. As the lowest detectable effect, transient reductions in lung function are mentioned, and because these have been shown to persist for some weeks after the episodes that were studied they have been graded as a moderate response. An increase in mortality has been identified as the most severe effect. The Working Group realized that there is continuing discussion about the existence as well as the magnitude of a threshold for effects of winter-type smog on mortality. It was decided that there was no compelling new evidence to refute the conclusions of the most recent WHO evaluation, and the levels given in Table 2 for this effect, 500 $\mu g/m^3$ sulfur dioxide and 500 $\mu g/m^3$ black smoke as 24-hour averages, have been taken from *Air quality guidelines for Europe (8)*. Effects on morbidity were graded as moderate when they begin to occur. The Group felt that these effects become severe at some point before effects on mortality begin to occur. In the absence of data obtained in the relevant range of exposure, the Working Group arbitrarily selected 24-hour average levels of 400 $\mu g/m^3$ sulfur dioxide combined with 400 $\mu g/m^3$ particulate matter in air as the threshold for severe effects.

The expected acute effects of summer-type smog are graded in Table 3. The effects on symptoms and lung function expected at summer-type smog exposures characterized by

Table 2. Levels of 24-hour average concentrations of air pollutant mixtures containing sulfur dioxide and particulate matter above which specific acute effects on human health are expected on the basis of observations made in epidemiological studies

Concentration ($\mu g/m^3$) of:		Health effects	Overall classification
sulfur dioxide	particulate matter		
200	200 (gravimetric)	Small, transient decrements in lung function (FVC, FEV_1) in children and adults that may last for 2 – 4 weeks. The magnitude of the effect is in the order of 2 – 4% of the group mean	Moderate
250	250 (black smoke)	Increase in respiratory morbidity among susceptible adults (chronic bronchitics) and possibly children	Moderate
400	400 (black smoke)	Further increase in respiratory morbidity	Severe
500	500 (black smoke)	Increase in mortality among elderly, chronically ill people	Severe

one-hour average concentrations of ozone of about 200 $\mu g/m^3$ were graded as mild. On the other end of the scale, the combination and intensity of effects expected at summer-type smog exposures characterized by one-hour average concentrations of ozone of about 400 $\mu g/m^3$ and over were graded as severe.

These levels do not indicate thresholds of effects, but indicate an amount of air pollution high enough to cause effects that may be detected in well designed studies. Higher levels of exposure will cause effects of increasing severity in an increasing fraction of the populations exposed; however, it is not possible to define this increase precisely on the basis of the limited data now available. A level of pollution lower than the lowest in the tables is not thought to be without effect, but is not expected to cause effects of major health concern. In general, those with pre-existing lung disease or circulatory deficiencies are more severely impaired than

Table 3. Expected acute effects of photochemical smog on days characterized by maximum 1-hour average ozone concentrations, as indicated for children and non-smoking young adults on the basis of observations made in toxicological, clinical and epidemiological studies

Ozone level ($\mu g/m^3$)	Eye, nose and throat irritation	Average FEV_1 decrement in active people outdoors		Imposed avoidance of time and activity outdoors	Respiratory inflammatory and clearance response, hyper-reactivity in active people outdoors	Respiratory symptoms (mainly in adults)	Overall classification
		Whole population	Most sensitive 10% of population				
< 100	No effect	None	None	None	None	None	—
200	In few sensitive people	5%	10%	None	Mild	Some chest tightness, cough	Mild
300	< 30% of people	15%	30%	Some individuals	Moderate	Increased symptoms	Moderate
400	> 50% of people	25%	50%	Many individuals	Severe	Further increase of symptoms	Severe

Note. In large cities, scavenging of ozone may lead to relatively low concentrations of ozone. Under such circumstances, other indicators of summer-type smog may be more useful.

others even by relatively small effects brought about by winter-type episodes. For summer-type smogs, people at special risk have not been clearly defined, although it is well known that some are more responsive to summer-type smog and to ozone in particular.

Measures to reduce health risks

To reduce the risks to health of winter- and/or summer-type smog exposures, measures need to be taken that primarily aim at reducing personal exposure to the smog mixture (or, rather, the delivered dose of pollution to the target organ). Among these are short- and long-term emission reduction measures, and measures to induce behavioural changes that minimize personal exposure or delivered dose. Below, measures to be taken are discussed in general with the understanding that, under specific local circumstances, specific measures are required.

Peak exposures to winter- or summer-type smog can be reduced when action taken in response to an air pollution "alert" results in reduced emissions and/or reduction of exposure through restrictions on personal mobility. Such alerts can be invoked only a limited number of times in any one season or year, and do not result in any substantial reduction in cumulative or long-term average exposures. The official response to a prediction that alert levels in a given location or jurisdiction will be exceeded is the responsibility of the recognized authority. When effects are expected to be mild, no other action than announcement of the expected alert and its public health significance seems necessary. When effects are expected to be moderate, some public advice about exposure or dose reduction for sensitive individuals could be considered. When severe health effects are expected, additional measures can be recommended on a voluntary basis, and emergency short-term measures such as the closing of schools or limiting of traffic can be considered.

The usefulness of short-term reductions in emissions by industry or traffic has been the subject of intense debate in several European countries over the past few years. In view of the economic cost and other disadvantages to society, authorities have been reluctant to introduce large-scale traffic bans during episodes of smog. Limited experience of traffic bans enforced in winter-type smog episodes in Germany has shown that these lead to extreme overloading of the public transport system. As a result, outdoor exposure to the air pollution mix tends to increase as people wait for buses and trains, walk to stations and bus stops, or decide to walk or bicycle to work. As a consequence, the delivered dose on a population basis may increase rather than decrease, even when short-term emission reduction measures succeed in reducing peak concentrations of indicator pollutants in the air. In summer-type smog this is an even more pressing problem, as these episodes are associated with warm, sunny weather that encourages people to spend time outdoors.

In the past, space heating was a major source of air pollution in winter in areas such as London. In some areas of eastern Europe this is still the case, and reduced heating is a possible way of lowering pollution emissions. However, authorities should be reluctant to advise people to reduce the heating of their homes, as exposure to low temperatures carries health risks of its own. A WHO working group convened in 1985 discussed the health risks associated with exposure to low temperatures in the home *(45)*. The conclusions were that there are no demonstrable risks to the health of healthy, sedentary people at air temperatures of between 18 °C and 24 °C. According to the report, there is evidence that ambient air temperatures below 12 °C are a health risk to elderly, sick or handicapped people, and to preschool children. For certain groups such as the sick, the handicapped, the very old and the very young, a minimum temperature of 20 °C was recommended.

Reduction of time spent outdoors and/or of physical activity outdoors is another way to reduce the risks to health of exposure to smog. In winter, being indoors protects to a certain extent from pollutants in outdoor air, although pollutants such as fine particulate matter may penetrate relatively easily. In summer, homes are generally better ventilated than in winter, so that being indoors offers less protection. However, even under well ventilated circumstances indoor ozone concentrations tend to be clearly lower than ambient concentrations due to the high reactivity of ozone. The inhaled dose per unit time of a pollutant can easily be more than five times higher during exercise than in a person at rest. Reduction of physical activity exerted outdoors is, therefore, sound advice to those seeking to reduce the adverse effects that exposure to either type of smog may have on their health. Chamber experiments with ozone have dramatically illustrated the effect of physical exercise on the severity of the health effects observed in these studies. This effect has not been demonstrated for winter smog.

Model calculations such as those of de Leeuw *(12)* have shown that in summer-type smog, elimination of local traffic emissions has a rather limited effect on peak ozone concentrations when such measures are taken during the episode only. Long-term reduction of baseline emissions of the relevant pollutants or precursors seems called for if exposure during smog episodes, as well as at all other times, is to be significantly reduced.

The need to convene the Working Group was unanimously felt to be evidence of the inadequacy of air pollution abatement measures, and the results of the meeting should in no way distract the responsible authorities from the need to increase efforts to reduce baseline levels. With reduced baseline emissions, the peak concentrations resulting from variable power demands and meteorology will also be reduced. Reductions in peak concentrations by this approach avoid emergency restrictions on economic and personal activity, and have the added advantage of reducing long-term cumulative exposures to smog pollutants. Data presented to the Working Group from various countries show that alert systems have not universally been adopted. Switzerland, for example, has chosen not to use an alert system for summer-type smog because short-term emergency measures would absorb resources required to realize medium-term measures aimed at permanently reducing precursor emissions. Neighbouring Austria, on the contrary, has created a warning system for summer-type smog consisting of a pre-warning and two warning levels of ozone. At the highest warning level of 400 $\mu g/m^3$ as a 3-hour mean, school sports and other strenuous outdoor activities may be forbidden by the responsible authorities.

If an alert system is adopted, it should specify exactly how one determines that a certain alert level has been or will be exceeded. This includes specifying such factors as the number and location of measuring stations at which the alert level is exceeded or is expected to be exceeded, and the prediction model.

Risk communication

Recent experience in several European countries has shown that air pollution episodes have the potential to draw intensive media attention and, as a consequence, to create great concern among the general public that may not be justified by the severity of the anticipated effects on health *(46)*. The psychological effects and their implications for human well-being may have been greatly underestimated. In particular, advice to the public with a strong impact on the normal activity pattern (such as advice to stop outdoor physical activity and/or not to leave the house) is likely to be associated with perceptions of emergency by at least some of the population. Effective communication of the level of health risk associated with exposure to different levels of air pollution, and of the level of air pollution at any given time, is critically important in maintaining the necessary confidence and cooperation of the public. This calls for a thorough education of the general population.

There are five specific points to which attention should be paid *(46)*.

- *The pollution level* at which information should be made available to the public.
- *The timing of the information*. This can be prospective, as a forecast for up to 48 hours; current, informing the public as to the present concentration; or retrospective, informing the public about the state of affairs in the preceding hours or days.

- *The target group*, such as the physically active general population in the case of summer-type smog, or the respiratory or cardiovascular disease patient in the case of winter-type smog.
- *The channel of information*. This could vary from using professionals working in the health care system as informers of the public, to widespread media attention through radio and television.
- *The wording of the information*. This could vary from deliberately rather vague characterizations such as "good" or "bad" air quality, to very specific recommendations on what to do and what not to do during smog episodes.

Conclusions and recommendations

Conclusions

1. While it is possible to select certain "index" pollutants characterizing the two main types of smog under consideration – such as sulfur dioxide and particulate matter in the case of winter-type smog and ozone in the case of summer-type – there can be substantial variations in the composition of the pollution mixture between different localities, and findings on effects in any one place will not necessarily apply elsewhere.

2. Although controls on emissions and changes in fuels have eliminated the severe health problems that used to occur in association with winter-type smog episodes, there are places where, due to inadequate emission control coupled with meteorological and topographical features, high pollution episodes of this type still occur, leading to acute effects on health.

3. Summer-type smog episodes have begun to occur in many areas of Europe over the past 20 years, their intensity and frequency tending to increase rather than decrease, and on the basis of experience elsewhere, acute effects on health are anticipated.

4. Peak exposures to winter- or summer-type smog can be reduced when action taken in response to an air-pollution alert results in reduced emissions from economic activity and/or

reductions of exposure through restrictions on personal mobility. However, the much preferred alternative to alert systems is to reduce peak exposure by reducing the baseline exposure using approaches such as fuel-switching, process changes and/or emission controls that lower baseline rates of emission in large enough areas.

5. Alert systems in respect of winter-type smog may need to be based on some combination of sulfur dioxide and particulate matter. In some circumstances it may be appropriate to use an excess of just one of these as an indicator, but otherwise various combinations might be used, depending on local circumstances and the corresponding evaluation of the health risks.

6. In view of the different levels used to distinguish the various health effects of winter- and summer-type smog episodes, the level of index pollutants should be determined with a margin of error of no more than ± 15%. In principle, the number of stations required to meet the demand for accuracy will depend on the spatial gradient over the area under consideration.

Recommendations

1. In areas subject to smog episodes of either type, monitoring should be extended to a wider range of pollutants than the basic "index" ones, so as to characterize the mixture; where possible, epidemiological studies should be undertaken locally to provide information for health risk assessment.

2. Authorities should take action to educate the general public about the potential acute health consequences of winter- and summer-type smog episodes in a way commensurate with the seriousness of the problem, and about the appropriate action for individuals to take to protect themselves against adverse effects.

3. With increasing levels of exposure during smog episodes, specific advice to the public should be considered about exposure or dose reduction for sensitive individuals.

4. Short-term source-reduction strategies that require the continuous cooperation of a large number of individuals, or which impose restrictions on their free choice, should be relied on only in severe smog episodes. When this kind of strategy is adopted, the budget should be sufficient to evaluate the effectiveness of the strategy in terms of a real reduction in health risk.

5. Measuring stations used to evaluate smog episodes should be sited in such a way that the results are representative of the exposure sustained by the population under consideration. Measurement results should be available on-line, with averaging times of three hours at the most, to make possible extrapolation to the next 24 or 48 hours.

6. To increase the quality of smog prognoses a simple model should be used, in which some characteristic meteorological factors are combined with measured pollution levels.

7. Professionals working in the health care system and other relevant staff should be provided with enough information to enable them to give advice to anybody concerned about the meaning of the health effects described in Tables 2 and 3.

References

1. Lübkert, B. Characteristics of the mid-January 1985 SO_2 smog episode in central Europe. Report from an international workshop. *Atmospheric environment*, **23**: 611 – 623 (1989).
2. Lahman, E. *Winter smog in Germany*. Berlin, WHO Collaborating Centre for Air Quality Management and Air Pollution Control, 1990 (Report 3/90).
3. Wichmann, H.E. et al. Health effects during a smog episode in West Germany in 1985. *Environmental health perspectives*, **79**: 89 – 100 (1989).
4. Yocom, J.E. Indoor–outdoor air quality relationships: a critical review. *Journal of the Air Pollution Control Association*, **32**: 500 – 520 (1982).
5. Hoek, G. et al. Indoor exposure to airborne particles and nitrogen dioxide during an air pollution episode. *Journal of the Air Pollution Control Association*, **39**: 1348 – 1349 (1989).
6. Commins, B.T. & Waller, R.E. Observations from a ten year study of pollution at a site in the city of London. *Atmospheric environment*, **1**: 49 – 68 (1967).
7. Onderdelinden, D. et al. *Luchtverontrenigingsepisode van 14 – 24 januari 1987, concentraties van SO_2 en NO_2 in Nederland en nabije omgeving* [Air pollution episode from 14 – 24 January 1987, concentrations of SO_2 and NO_2 in the Netherlands and immediate surroundings]. Bilthoven, National Institute of Public Health and Environmental Hygiene, 1987 (Report 228702001).
8. *Air quality guidelines for Europe*. Copenhagen, WHO Regional Office for Europe, 1987 (WHO Regional Publications, European Series, No. 23).

9. Seinfeld, J.H. Urban air pollution: state of the science. *Science*, **243**: 745 – 752 (1989).
10. Weschler, C.J. et al. Indoor ozone exposure. *Journal of the Air Pollution Control Association*, **39**: 1562 – 1568 (1989).
11. Hayes, S.R.F. Estimating the effect of being indoors on total personal exposure to outdoor air pollution. *Journal of the Air Pollution Control Association*, **39**: 1453 – 1461 (1989).
12. de Leeuw, F.A.A.M. Contributions of traffic emissions to mesoscale NO_x and O_3 concentrations. *Atmospheric environment*, **23**: 49 – 53 (1989).
13. Hatzakis, A. et al. Short-term effects of air pollution on mortality in Athens. *International journal of epidemiology*, **15**: 73 – 81 (1986).
14. Katsouyanni, K. et al. Air pollution and cause specific mortality in Athens. *Journal of epidemiology and community health*, **44**: 321 – 324 (1990).
15. Derriennic, F. et al. Short-term effects of sulphur dioxide pollution on mortality in two French cities. *International journal of epidemiology*, **18**: 186 – 197 (1989).
16. Lawther, P.J. et al. Air pollution and exacerbations of bronchitis. *Thorax*, **25**: 525 – 539 (1970).
17. Englert, N. *Air quality and respiratory diseases in children in Berlin (West)*. Berlin, WHO Collaborating Centre for Air Quality Management and Air Pollution Control, 1990 (Report 1/90).
18. Samet, J.M. et al. The relationship between air pollution and emergency room visits in an industrial community. *Journal of the Air Pollution Control Association*, **31**: 236 – 240 (1981).
19. van der Lende, R. et al. A temporary decrease in the ventilatory function of an urban population during an acute increase in air pollution. *Bulletin européen de physiopathologie respiratoire*, **11**: 31 – 43 (1975).
20. Stebbings, J.H., Jr. & Fogleman, D.G. Identifying a susceptible subgroup: effects of the Pittsburgh air pollution episode upon schoolchildren. *American journal of epidemiology*, **110**: 27 – 40 (1979).
21. Dockery, D.W. et al. Change in pulmonary function in children associated with air pollution episodes. *Journal of the Air Pollution Control Association*, **32**: 937 – 942 (1982).
22. Brunekreef, B. et al. Sensitive subgroups and normal variation in pulmonary function response to air pollution episodes. *Environmental health perspectives*, **90**: 189 – 193 (1991).

23. DASSEN, W. ET AL. Decline in children's pulmonary function during an air pollution episode. *Journal of the Air Pollution Control Association*, **36**: 1223 – 1227 (1986).
24. WICHMANN, H.E. ET AL. Lungenfunktion und Carboxyhämoglobin in der Smogsituation des Januar 1987 (Pulmonary function and carboxyhemoglobin during the smog episode in January 1987). *Zentralblatt für Bakteriologie, Mikrobiologie und Hygiene, Series B*, **187**: 31 – 43 (1988).
25. BRUNEKREEF, B. ET AL. Pulmonary function changes associated with an air pollution episode in January 1987. *Journal of the Air Pollution Control Association*, **39**: 1444 – 1447 (1989).
26. SPENGLER, J.D. ET AL. Acid air and health. *Environmental science and technology*, **24**: 946 – 956 (1990).
27. LIPPMANN, M. Health effects of ozone: a critical review. *Journal of the Air Pollution Control Association*, **39**: 672 – 695 (1989).
28. ROMBOUT, P.J.A. ET AL. Rationale for an eight-hour ozone standard. *Journal of the Air Pollution Control Association*, **36**: 913 – 917 (1986).
29. AVOL, E.L. ET AL. Comparative respiratory effects of ozone and ambient oxidant pollution exposures during heavy exercise. *Journal of the Air Pollution Control Association*, **34**: 804 – 809 (1984).
30. HIGGINS, I.T.T. ET AL. Effect of exposures to ambient ozone on ventilatory lung function in children. *American review of respiratory disease*, **141**: 1136 – 1146 (1990).
31. SPEKTOR, D.M. ET AL. Effects of ambient ozone on respiratory function in healthy adults exercising outdoors. *American review of respiratory disease*, **138**: 821 – 828 (1988).
32. SPEKTOR, D.M. ET AL. Effects of ambient ozone on respiratory function in active, normal children. *American review of respiratory disease*, **137**: 313 – 320 (1988).
33. FOLINSBEE, L.J. ET AL. Pulmonary function and symptom responses after 6.6 hour exposure to 0.12 ppm ozone with moderate exercise. *Journal of the Air Pollution Control Association*, **38**: 28 – 35 (1988).
34. HORSTMAN, D.H. ET AL. Ozone concentration and pulmonary response relationships for 6.6 hour exposures with five hours of moderate exercise to 0.08, 0.10 and 0.12 ppm. *American review of respiratory disease*, **142**: 1158 – 1163 (1990).
35. KOENIG, J.Q. ET AL. Acute effects of 0.12 ppm ozone or 0.12 ppm nitrogen dioxide on pulmonary function in healthy and asthmatic adolescents. *American review of respiratory disease*, **132**: 648 – 651 (1985).

36. Ostro, B.D. et al. Predicting respiratory morbidity from pulmonary function test: a reanalysis of ozone chamber studies. *Journal of the Air Pollution Control Association*, **39**: 1313 – 1318 (1989).
37. Schwartz, J. & Zeger, S. Passive smoking, air pollution and acute respiratory symptoms in a diary study of student nurses. *American review of respiratory disease*, **141**: 62 – 67 (1990).
38. Whittemore, A.S. & Korn, E.L. Asthma and air pollution in the Los Angeles Area. *American journal of public health*, **70**: 687 – 681 (1980).
39. Bates, D.V. & Sizto, R. Air pollution and hospital admissions in southern Ontario: the acid summer haze effect. *Environmental research*, **43**: 317 – 331 (1987).
40. Koenig, J.Q. Prior exposure to ozone potentiates subsequent response to sulphur dioxide in adolescent asthmatic subjects. *American review of respiratory disease*, **141**: 377 – 380 (1990).
41. Koren, H.S. et al. Ozone-induced inflammation in the lower airways of human subjects. *American review of respiratory disease*, **139**: 407 – 415 (1989).
42. Koren, H.S. et al. The inflammatory response in human lung exposed to ambient levels of ozone. *In*: Schneider, T. et al., ed. *Atmospheric ozone and its policy implications*. Amsterdam, Elsevier, 1989 (Studies in Environmental Science 35), pp. 745 – 753.
43. Lippmann, M. Health significance of pulmonary function responses to airborne irritants. *Journal of the Air Pollution Control Association*, **38**: 881 – 889 (1988).
44. American Thoracic Society. Guidelines as to what constitutes an adverse respiratory health effect, with special reference to epidemiologic studies of air pollution. *American review of respiratory disease*, **131**: 666 – 668 (1985).
45. *Health impact of low indoor temperatures*. Copenhagen, WHO Regional Office for Europe, 1987 (Environmental Health Series 16).
46. Wagner, H.M. *Photochemical smog in Europe – an overview*. Berlin, WHO Collaborating Centre for Air Quality Management and Air Pollution Control, 1990 (Report 2/90).

Annex 1

Membership of subgroups

Subgroup 1 on summer-type smog

Dr L. van Bree
Dr A. Hasse
Dr K. Katsouyanni
Dr E. Lebret
Dr R. Maynard
Dr D. Menzel
Dr P.J.A. Rombout
Dr J.A.J. Stolwijk (*Rapporteur*)
Dr M. Wagner (*Leader*)
Dr H.-U. Wanner

Subgroup 2 on winter-type smog

Dr T.P.M. Blom
Dr B. Brunekreef (*Rapporteur*)
Dr N. Englert
Dr M. Haider
Dr M. Lippmann (*Leader*)
Dr D. Onderdelinden
Dr J.A. Sokal
Mr R.E. Waller

Annex 2

Participants

Temporary advisers

Mr Ton P.M. Blom, Staff Officer, Air Pollution Directorate, Ministry of Housing, Physical Planning and Environment, Leidschendam, Netherlands

Dr Leendert van Bree, Senior Toxicologist, Department of Inhalation Toxicology, National Institute of Public Health and Environmental Hygiene, Bilthoven, Netherlands

Dr Bert Brunekreef, Associate Professor, Department of Environmental and Tropical Health, Wageningen Agricultural University, Wageningen, Netherlands (*Rapporteur*)

Dr Norbert Englert, Medical Officer, Institute of Water, Soil and Air Hygiene, Federal Health Office, Berlin, Federal Republic of Germany

Dr Manfred Haider, Professor and Head, Institute of Environmental Hygiene, University of Vienna, Austria

Dr Arnim Hasse, Head, Section for Environmental Health Effects, Federal Environmental Agency, Berlin, Federal Republic of Germany

Dr Klea Katsouyanni, Assistant Professor, Department of Hygiene and Epidemiology, School of Medicine, University of Athens, Greece

Dr Erik Lebret, Coordinator, Environmental Epidemiology, Department of Epidemiology, National Institute of Public Health and Environmental Hygiene, Bilthoven, Netherlands

Dr Morton Lippmann, Professor, Institute of Environmental Medicine, New York University Medical Center, Tuxedo, USA (*Chairman*)

Dr Robert L. Maynard, Medical Officer, Toxicology and Environmental Health, Department of Health, London, United Kingdom

Dr Daniel Menzel, Professor and Chairman, Department of Community and Environmental Medicine, University of California, Irvine, USA

Dr Dirk Onderdelinden, Deputy Head, Laboratory for Air Research, National Institute of Public Health and Environmental Hygiene, Bilthoven, Netherlands

Dr Peter J.A. Rombout, Head, Department of Inhalation Toxicology, National Institute of Public Health and Environmental Hygiene, Bilthoven, Netherlands

Dr Jerzy A. Sokal, Head, Department of Toxicity Evaluation, Institute of Occupational Medicine, Lodz, Poland

Dr Jan A.J. Stolwijk, Professor, Department of Epidemiology and Public Health, Yale University School of Medicine, New Haven, CT, USA

Dr Michael Wagner, Professor of Environmental Hygiene, Technical University, Berlin, and Head, Laboratory of Air Pollution Effects, Air Hygiene Department, Institute of Water, Soil and Air Hygiene, Federal Health Office, Berlin, Federal Republic of Germany (*Vice-chairman*)

Mr Robert E. Waller, Principal Scientific Officer, Toxicology and Environmental Health, Department of Health, London, United Kingdom

Dr Hans-Urs Wanner, Professor, Institute of Hygiene and Work Physiology, Swiss Federal Technical University, Zurich, Switzerland

WHO Regional Office for Europe

Dr Michael J. Suess, Regional Officer for Environmental Health Hazards (*Scientific Secretary*)

Résumé

La réunion avait pour objet d'évaluer les risques de santé liés à l'exposition épisodique à de fortes concentrations de polluants atmosphériques en hiver et en été (smog de type «estival», ou de type «hivernal»). Elle était organisée avec la coopération et le soutien du Ministère néerlandais du logement, de l'aménagement du territoire et de l'environnement. Quinze conseillers temporaires et trois observateurs, venant de huit pays, y ont assisté.

Les participants ont examiné les effets sur la santé humaine connus ou présumés de l'exposition épisodique au smog d'hiver et au smog d'été. Aux fins du présent rapport, on entendra par smog d'hiver une forme de pollution provenant principalement de la combustion de combustibles fossiles contenant du soufre. Le dioxyde de soufre et les particules en suspension sont habituellement pris comme polluants témoins pour ce mélange, même si d'autres composantes de ce dernier telles que l'acide sulfurique peuvent être les causes premières d'effets de santé. Le smog d'été, par contre, est une forme de pollution photochimique résultant de réactions atmosphériques entre les hydrocarbures et les oxydes d'azote, stimulées par un intense rayonnement solaire. Dans ce mélange, l'ozone est considéré comme le polluant biologiquement le plus actif.

On ne doit pas perdre de vue cependant qu'à cause des larges variations de composition du smog d'hiver et du smog d'été d'un site à l'autre les constatations faites sur les effets en un lieu ne seront pas nécessairement valables ailleurs. Ainsi, si l'on compare des lieux qui sont situés à différentes latitudes, et donc soumis à un rayonnement solaire ultraviolet d'intensité différente, ou qui ont une contribution relative différente des sources, la toxicité du smog photochimique pour une même concentration d'ozone variera notablement. Des incertitudes existent aussi quant aux caractéristiques de l'exposition aux polluants atmosphériques et aux effets de santé associés au centre des grandes villes à forte circulation automobile; il serait urgent de les réduire.

Dans les tableaux de l'Annexe 1 sont présentées sous forme sommaire quelques informations sur les effets prévisibles en cas d'épisode de smog d'hiver et de smog d'été. En fonction du degré d'exposition, ces effets sont classés «faibles», «modérés» ou «sérieux», ordre qui exprime aussi bien entendu le degré d'urgence des mesures.

Pour réduire l'exposition de pointe au smog hivernal ou estival, on peut mettre en place un système d'alerte à la pollution atmosphérique, basé sur la réduction temporaire des émissions provenant de l'activité économique, et/ou la réduction temporaire de l'exposition par des mesures restreignant les déplacements personnels. Ces mesures sont appliquées par l'autorité compétente, lorsque la concentration d'un polluant témoin risque de dépasser une valeur prédéterminée jugée comme ayant des effets nuisibles pour la santé. Ces alertes, en pratique, peuvent seulement être déclenchées quelques fois par saison ou par an, et elles n'entraînent aucune réduction notable de l'exposition cumulative ou moyenne à long terme.

La réponse officielle à la prévision que le niveau d'alerte va être dépassé en un lieu donné ou dans une zone administrative donnée est décidée par les autorités compétentes. Lorsque les effets prévisibles sont faibles, il semble suffisant d'annoncer que les niveaux d'alerte risquent d'être dépassés et d'indiquer la signification de cet état de chose en matière de santé publique. Si les effets prévus atteignent le niveau «modéré», on peut envisager de donner au public certains conseils sur la manière de réduire l'exposition ou la dose pour les individus fragiles. Lorsque l'on prévoit des effets sérieux, les autorités peuvent recommander des mesures additionnelles à titre non obligatoire. La prise de mesures d'urgence de courte durée, telles que la fermeture des écoles ou les restrictions à la circulation, devrait être envisagée lorsque l'on s'attend à ce que les concentrations dépassent les valeurs causant des effets sérieux sur la santé humaine (tableaux 1 et 2 de l'Annexe 1).

Les valeurs indiquées dans les tableaux ne sont pas des valeurs seuils d'effet, mais correspondent à un degré de pollution suffisamment élevé pour causer des effets qui peuvent être détectés lors d'études sérieusement exécutées. A un niveau d'exposition encore supérieur, il y aura des effets de plus en plus sérieux dans une fraction de plus en plus nombreuse de la population exposée. Il n'est pas possible cependant d'attribuer à cet accroissement des valeurs précises étant donné le peu de données dont on dispose actuellement. Inversement, lorsque le niveau de pollution est inférieur à la valeur la plus basse indiquée dans les tableaux, cela ne veut pas dire, comme on l'a indiqué plus haut, que cette pollution soit inoffensive, mais qu'elle n'est pas jugée susceptible de causer des effets sérieux en matière de santé.

En général, les personnes déjà atteintes d'insuffisance pulmonaire ou circulatoire sont plus fortement affectées que le reste de la population par des épisodes de smog d'hiver même relativement bénins. En ce qui concerne le smog d'été, les personnes à haut risque n'ont pas encore été déterminées avec précision; il est toutefois bien connu que certains individus réagissent plus fortement à cette pollution, et en particulier à celle due à l'ozone.

Le groupe dans son ensemble a estimé que la tenue même d'une réunion ayant pour objet d'évaluer les risques pour la santé résultant des épisodes de smog semblait indiquer l'inefficacité des mesures de réduction de la pollution de l'air. Quels que soient les résultats auxquels elle aboutirait, les autorités responsables ne devraient en aucun cas relâcher leurs efforts pour réduire les niveaux moyens de pollution. Cette réduction en effet ferait automatiquement baisser les concentrations de pointe résultant des variations de la demande d'électricité et des conditions climatiques, ce qui rendrait superflues toutes les mesures d'urgence restreignant l'activité économique ou les déplacements personnels, en plus des gains obtenus en ce qui concerne l'exposition cumulative à long terme. (A ce propos, on pourra consulter le rapport sur l'*Impact on human health of air pollution in Europe*, établi par le Bureau régional de l'OMS pour l'Europe pour le compte de la Commission économique pour l'Europe des Nations Unies à Genève[a].)

L'exemple récent de plusieurs pays européens montre que les incidents de pollution de l'air pouvaient mobiliser l'attention des médias et, par voie de conséquence, susciter dans le public une inquiétude souvent hors de proportion avec les effets prévisibles sur la santé humaine. Cette réaction prouve en tous cas que l'on a jusqu'ici grandement sous-estimé les effets psychologiques de ces incidents et leur impact sur le bien-être. En particulier, les conseils au public ayant des répercussions directes sur les activités normales de celui-ci (conseils d'interrompre les activités physiques en plein air ou de rester chez soi par exemple) risquent d'être interprétés comme indiquant une situation d'urgence par une partie au moins de la population. C'est pourquoi une parfaite communication en ce qui concerne le degré de risque pour la santé associé à l'exposition à différents taux de pollution atmosphérique, et le niveau de la pollution atmosphérique à tout moment, est d'une importance primordiale si l'on veut entretenir le climat indispensable de confiance et de coopération de la part du public. Il sera donc essentiel d'entreprendre un programme sérieux d'éducation à l'intention du grand public.

Conclusions

1. Bien qu'il soit en principe possible de choisir certains polluants comme témoins pour les deux types de smog qui nous intéressent, à savoir le dioxyde de soufre et les particules en suspension pour le smog hivernal et l'ozone pour le smog estival, il existe des variations importantes de la composition du mélange polluant d'un endroit à l'autre, et les constatations sur les effets en un endroit donné ne s'appliqueront pas nécessairement ailleurs.

2. Bien que les mesures prises pour lutter contre les émissions, notamment par utilisation d'autres combustibles, aient le plus souvent éliminé les graves problèmes de santé autrefois causés par les épisodes de smog hivernal, il subsiste des endroits où l'insuffisance des mesures antipollution, associée aux conditions

[a] Ce document peut être obtenu à l'unité Elimination des risques pour la santé liés à l'environnement, Bureau régional de l'OMS pour l'Europe, Scherfigsvej 8, DK-2100 Copenhague Ø.

météorologiques et topographiques locales, favorise des épisodes de forte pollution, ayant des effets aigus sur la santé.

3. En ce qui concerne le smog de type estival, des épisodes se sont produits dans de nombreux endroits d'Europe au cours des vingt dernières années; leur intensité et leur fréquence tendent plutôt à augmenter et, compte tenu de l'expérience acquise ailleurs, on peut s'attendre à ce qu'ils entraînent des effets aigus en matière de santé.

4. Il est certes possible de réduire l'exposition aux concentrations de pointe du smog hivernal ou estival en instaurant un système d'alerte à la pollution de l'air accompagné de mesures restreignant l'activité économique pour réduire les émissions, ou les déplacements personnels pour réduire l'exposition. Il est cependant bien préférable de faire baisser l'exposition de pointe en réduisant l'exposition moyenne grâce à l'utilisation d'autres combustibles, à la modification des procédés ou à des mesures de limitation des émissions ayant un effet sur des zones suffisamment vastes.

5. En ce qui concerne les systèmes d'alerte pour le smog de type hivernal, en dehors des cas exceptionnels où l'on pourra se baser sur le dépassement d'un seul indicateur, il sera en général nécessaire de prendre comme critère une combinaison des concentrations de dioxyde de soufre et de particules en suspension, qui variera selon les conditions locales et l'évaluation des risques de santé résultants.

6. En ce qui concerne les différentes valeurs appliquées pour distinguer entre les divers degrés d'effets de santé des épisodes de smog hivernal et estival, une précision de l'ordre de 15% semble nécessaire pour la détermination des concentrations réelles de polluants témoins. En principe, le nombre de stations nécessaires pour répondre à cette condition sera fonction de la répartition topographique dans la zone considérée.

Recommandations

1. Dans les zones exposées à des épisodes de smog hivernal ou estival, la surveillance devrait être étendue à d'autres polluants

que les quelques polluants témoins, pour permettre de définir précisément le mélange. Lorsqu'il est possible, on devrait exécuter des enquêtes épidémiologiques au niveau local afin de disposer d'informations pour l'évaluation des risques de santé.

2. Les autorités doivent entreprendre d'informer le grand public des effets de santé aigus potentiels des épisodes de smog hivernal et estival, d'une manière adaptée à la gravité du problème, ainsi que des mesures que l'individu peut prendre pour se protéger contre les effets nocifs de ces épisodes.

3. A partir d'un certain degré d'exposition en cas d'épisodes de smog, il faudrait envisager de donner au public des conseils précis sur la manière de réduire l'exposition ou de réduire la dose pour les individus vulnérables.

4. Les mesures de courte durée de réduction à la source nécessitant une coopération continue d'un grand nombre de personnes, ou imposant des restrictions à la liberté de déplacement de nombreuses personnes, devraient être réservées aux cas graves de smog. Si l'on utilise des mesures de cette nature, il faudra prévoir au budget des crédits suffisants pour que l'on puisse évaluer leur efficacité réelle du point de vue de la réduction des risques de santé.

5. Les stations utilisées pour mesurer les concentrations lors d'épisodes de smog doivent être situées de telle manière qu'elles donnent des résultats représentatifs de l'exposition réelle des populations concernées. Les résultats des mesures devraient pouvoir être obtenus en temps réel, avec une base de mesure moyenne de trois heures au maximum, pour permettre des extrapolations pour les prochaines 24 ou 48 heures.

6. Pour améliorer la qualité des pronostics relatifs au smog, il faudrait disposer d'un modèle simple établissant une relation entre certains facteurs météorologiques caractéristiques et les niveaux de pollution mesurés.

7. Les professionnels du système de santé et les autres personnels concernés devraient être suffisamment bien informés pour pouvoir eux-mêmes conseiller toute personne leur posant des questions sur les effets de santé décrits à l'Annexe 1 du présent document.

Annexe I

Effets de santé connus du smog de type hivernal et de type estival

Tableau I. Valeurs des concentrations moyennes sur 24 heures de SO_2 et de particules en suspension et effets correspondants sur la santé humaine, d'après les résultats d'enquêtes épidémiologiques

SO_2	Particules ($\mu g/m^3$)	Effets de santé	Degré d'effet
200	200 (mesure gravimétrique)	Faible réduction temporaire de la fonction pulmonaire (CV, VEMS) chez les enfants et les adultes, d'une durée de 2 à 4 semaines. L'ampleur des effets est de l'ordre de 2 – 4% de la moyenne pour le groupe.	modéré
250	250 (fumée noire)	Accroissement de la morbidité respiratoire chez les adultes fragiles (bronchite chronique), et éventuellement les enfants	modéré
500	500 (fumée noire)	Accroissement de la mortalité chez les personnes âgées et les malades chroniques	sérieux

Note : Les effets de santé chez l'homme sont considérés comme *sérieux* à partir de valeurs de 400 $\mu g/m^3$ aussi bien pour le dioxyde de soufre que pour les particules en suspension.

Tableau 2. Effets aigus estimés de l'exposition au smog photochimique pour des concentrations maximales journalières (moyenne sur une heure) d'ozone chez les enfants et les jeunes adultes non fumeurs, d'après les résultats d'enquêtes toxicologiques, cliniques et épidémiologiques (les effets chroniques ne sont pas considérés)

Concentration d'ozone ($\mu g/m^3$)	Irritation des yeux, du nez et de la gorge (par rapport à la population totale)	Diminution moyenne du VEMS chez les personnes actives (vivant à l'extérieur)		Restrictions au séjour et à l'activité en plein air	Réponse respiratoire, inflammatoire, clairance, hyperréactivité (chez les personnes actives séjournant à l'extérieur)	Symptômes respiratoires (principalement chez les adultes)	Degré d'effet
		Population totale	Fraction 10% la plus vulnérable				
< 100	Néant	Néant	Néant	Néant	Néant	Néant	
200	Chez un petit nombre de sujets vulnérables	5%	10%	Néant	Faible	Légère gêne respiratoire, toux	Faible
300	< 30% de la population	15%	< 30%	Quelques individus	Modéré	Symptômes aggravés	Modéré
400	> 50% de la population	25%	> 50%	De nombreuses personnes	Sérieux	Symptômes encore aggravés	Sérieux

Zusammenfassung

Ziel der Arbeitsgruppe war die Abschätzung der Gesundheitsrisiken durch episodisch auftretende höhere Luftschadstoffkonzentrationen im Winter und im Sommer (Wintersmog und Sommersmog). Veranstaltet wurde die Tagung zusammen mit dem niederländischen Ministerium für Wohnungswesen, Raumplanung und Umweltschutz; es nahmen fünfzehn Berater auf Zeit und drei Beobachter aus insgesamt acht Ländern teil.

Im Sinne des vorliegenden Kurzberichts wird Wintersmog hauptsächlich durch Verbrennung von schwefelhaltigen fossilen Brennstoffen hervorgerufen. Als Schadstoffanzeiger dieser Mischung werden gewöhnlich die Schwefeldioxid- und Schwebstoffkonzentrationen gemessen, obwohl andere Komponenten wie z.B. Schwefelsäure für die gesundheitlichen Auswirkungen hauptverantwortlich sein können. Für den Sommersmog typisch sind photochemische Umsetzungen von Kohlenwasserstoffen und Stickoxiden unter Einfluß des stärkeren Sonnenlichts. Bei dieser Mischung wird Ozon als biologisch aktivster Schadstoff betrachtet.

In der Zusammensetzung des Winter- und Sommersmogs kann es aber von Ort zu Ort erhebliche Schwankungen geben und die an einem Ort ermittelten Wirkungen brauchen anderenorts nicht zu gelten. Wenn z.B. Standorte auf verschiedenen Breitengraden mit verschiedener ultravioletter Sonnenstrahlung und mit verschiedenen Schadstoffquellen miteinander verglichen werden, ist die Toxizität des photochemischen Smogs bei der gleichen Ozonkonzentration wahrscheinlich sehr verschieden. Desgleichen muß baldigst die Unsicherheit in bezug auf die Merkmale der Exposition gegenüber Luftschadstoffen im

Zentrum großer Städte mit intensivem Kfz-Verkehr und den damit verbundenen gesundheitlichen Auswirkungen abgebaut werden.

Anhang 1 enthält Informationen über die gesundheitlichen Effekte, die im Zusammenhang mit typischen Winter- und Sommersmogepisoden erwartet werden. Die Exposition wurde in die drei Klassen „leichte", „mittlere" und „schwere" eingeteilt, für die verschiedene Gegenmaßnahmen gelten.

Die Spitzenbelastungen beim Winter- und Sommersmog können reduziert werden, wenn im Rahmen eines Smogalarms die industriellen und gewerblichen Emissionen eingeschränkt und/oder die menschliche Exposition durch Einschränkung der Mobilität vermindert wird. Ein Smogalarm wird definiert als offizielle Maßnahme der einschlägigen Behörde, die bei Überschreitung des Grenzwertes für einen Schadstoffanzeiger ausgelöst wird, wobei zuvor eine Relation zwischen dem Grenzwert und schädlichen Gesundheitswirkungen hergestellt worden ist. Alarme dieser Art können in der Praxis nur wenige Male in einer bestimmten Jahreszeit oder in einem Jahr ausgelöst werden und bewirken keine merkliche Senkung der kumulativen bzw. langfristigen Durchschnittsexposition.

Die offiziellen Maßnahmen bei voraussichtlicher Überschreitung der Grenzwerte in einem bestimmten Gebiet sind von der dafür zuständigen Behörde zu treffen. Wenn eine leichte Exposition erwartet wird, reicht es wahrscheinlich aus, die Bevölkerung darauf aufmerksam zu machen, daß die Grenzwerte möglicherweise überschritten werden, und der Bevölkerung zu erklären, was dies praktisch bedeutet. Wenn man mit einer mittleren Exposition rechnet, ist der Bevölkerung zu raten, daß anfällige Bürger bestimmte Verhaltensregeln befolgen sollten, um die Exposition bzw. Schadstoffdosis zu begrenzen. Bei voraussichtlich schwerer Exposition können zusätzliche Maßnahmen auf freiwilliger Basis empfohlen werden. Kurzfristige Notmaßnahmen wie Schließen der Schulen oder Einschränkung des Verkehrs sollten erwogen werden, wenn die Grenzwerte für schwere Expositionen voraussichtlich überschritten werden (Anhang 1, Tabelle 1 und 2).

Die in den Tabellen angegebenen Konzentrationen sind eigentlich keine Grenzwerte für bestimmte Effekte, sondern besagen nur, daß ab diesen Konzentrationen in gründlich durchgeführten Studien Effekte entdeckt werden können. Bei einer stärkeren Exposition nehmen die Auswirkungen in einem zunehmenden Teil der exponierten Personen zu; man kann diese Zunahme wegen der Knappheit der verfügbaren Daten nicht numerisch genau ausdrücken. Eine Schadstoffkonzentration unter dem niedrigsten Wert in den Tabellen hat unserer Auffassung nach aber auch eine Wirkung, bloß wird nicht erwartet, daß dies von größerer gesundheitlicher Bedeutung ist.

Im allgemeinen gilt, daß Personen, die bereits an einer Lungenkrankheit oder Kreislaufstörung leiden, schon bei geringen Schadstoffeinwirkungen in Verbindung mit dem Wintersmog stärker in Mitleidenschaft gezogen werden als andere. In bezug auf Sommersmog ist die spezielle Risikogruppe noch nicht eindeutig umrissen worden; es ist aber bekannt, daß manche Personen stärker darauf reagieren, besonders auf Ozon, als andere.

Die Gruppe war insgesamt der Meinung, daß die Tatsache, daß diese Tagung einberufen wurde, um die durch Smog verursachten Gesundheitsrisiken zu bewerten, beweise, daß die Maßnahmen zur Bekämpfung der Luftverschmutzung versagt hätten und die Tagungsergebnisse in keiner Weise bedeuteten, daß die zuständigen Behörden darauf verzichten könnten, sich um eine Senkung der Verschmutzungs-Basiswerte zu bemühen. Bei Senkung der Emissionsbasiswerte gehen auch die durch schwankenden Energieverbrauch und variierende Wetterverhältnisse bedingten Spitzenkonzentrationen zurück. Somit können Notmaßnahmen, die die wirtschaftlichen und persönlichen Tätigkeiten einschränken, vermieden werden; außerdem läßt sich dadurch die langfristige kumulative Smogschadstoffexposition reduzieren (In diesem Zusammenhang sei auf den Bericht „Impact on human health of air pollution in Europe" verwiesen, der vom europäischen WHO-Regionalbüro für die UN-Wirtschaftskommission für Europa, ECE, in Genf verfaßt wurde[a]).

In letzter Zeit hat es sich in mehreren europäischen Ländern gezeigt, daß die Medien sehr an Luftverschmutzungsfällen

---

[a] Zu beziehen vom Referat Bekämpfung umweltbedingter Gesundheitsgefahren, WHO-Regionalbüro für Europa, Scherfigsvej 8, DK-2100 Kopenhagen Ø.

interessiert sind und deshalb leicht die breite Öffentlichkeit verunsichern können; normalerweise rechtfertigen die zu erwartenden gesundheitlichen Auswirkungen nicht diese Reaktion. Die psychologische Wirkung und ihr Einfluß auf das Wohlbefinden sind aber wahrscheinlich stark unterschätzt worden. Insbesondere bewirken Ratschläge an die Bevölkerung, die stark in das tägliche Leben eingreifen (z.B. die Aufforderung, Tätigkeiten im Freien zu unterlassen bzw. nicht außer Haus zu gehen), zumindest in einem Teil der Bevölkerung, daß sie an einen akuten Notstand glaubt. Wenn man sich das Vertrauen der Öffentlichkeit und ihre Zusammenarbeitsbereitschaft sichern will, ist es von entscheidender Bedeutung, daß gründlich informiert wird über den Zusammenhang zwischen Gesundheitsgefahren und verschiedenen Luftschadstoffkonzentrationen sowie über die zu dem jeweiligen Zeitpunkt vorherrschende Schadstoffkonzentration. Das heißt, es muß eine intensive Aufklärung der breiten Öffentlichkeit betrieben werden.

Schlußfolgerungen

1. Zwar ist es möglich für die zwei Smogtypen Schadstoffanzeiger zu ermitteln, z.B. Schwefeldioxid und Schwebstoffe für Wintersmog und Ozon für Sommersmog, doch kann die Zusammensetzung des Schadstoffgemisches von Ort zu Ort sehr unterschiedlich sein, und die Resultate in bezug auf die gesundheitlichen Effekte lassen sich nicht an allen Orten in gleicher Weise anwenden.

2. Obwohl durch Einschränkung der Emissionen und Änderungen in bezug auf Kraft- bzw. Brennstoffe die ernsten Gesundheitsprobleme in Verbindung mit Wintersmog eingeschränkt worden sind, gibt es mancherorts infolge unzureichender Emissionsbekämpfung in Verbindung mit meteorologischen und topographischen Faktoren immer noch Verschmutzungsepisoden dieser Art mit akuten Gesundheitsbeeinträchtigungen.

3. Sommersmog ist in den letzten 20 Jahren in vielen europäischen Gebieten aufgetreten und zeigt in bezug auf Intensität und Frequenz eher eine zunehmende als eine abnehmende

Tendenz; ausgehend von anderenorts gemachten Erfahrungen wird mit zunehmenden gesundheitsschädlichen Effekten gerechnet.

4. Spitzenbelastungen bei Winter- und Sommersmog können reduziert werden, wenn bei Smogalarm die industriellen Emissionen reduziert und/oder die Exposition der Bevölkerung durch Einschränkung der persönlichen Mobilität gesenkt wird. Statt durch Warnmaßnahmen sollte man jedoch die Spitzenbelastungen durch Senkung der Verschmutzungs-Basiswerte in ausreichend großen Gebieten reduzieren, indem man auf andere Energiearten umschaltet, Arbeitsverfahren ändert und/oder die Emissionwerte senkt.

5. Warnsysteme für Wintersmog können auf den Meßwerten für Schwefeldioxid und Schwebstoffe basieren. Unter bestimmten Bedingungen ist es vorzuziehen, die Meßwerte nur für einen dieser Indikatoren zu benutzen; je nach den örtlichen Gegebenheiten und der entsprechenden Evaluation der Gesundheitsrisiken kann es ratsam sein, verschiedene Kombinationen zugrundezulegen.

6. Hinsichtlich der verschiedenen Konzentrationen zur Klassifizierung der Gesundheitseffekte von Winter- und Sommersmog wird davon ausgegangen, daß die Schadstoffanzeiger mit mindestens 85% Genauigkeit gemessen werden müssen. Im Prinzip richtet sich die Zahl der Meßstationen, wenn eine ausreichende Genauigkeit erzielt werden soll, nach der räumlichen Verteilung des Konzentrationsgefälles in dem zu untersuchenden Raum.

Empfehlungen

1. In Gebieten, in denen öfters Winter- oder Sommersmog vorkommt, sollte man nicht nur die Schadstoffanzeiger, sondern auch andere Schadstoffe überwachen, um die Zusammensetzung der Schadstoffmischung näher kennenzulernen; soweit möglich sollten örtlich begrenzte epidemiologische Studien durchgeführt werden, damit man Informationen über die Abschätzung der gesundheitlichen Risiken erhält.

2. Die Behörden sollten entsprechend dem Ernst der Lage die Öffentlichkeit über die potentiellen akuten Gesundheitsfolgen des Winter- und Sommersmogs sowie über die zu ergreifenden Schutzmaßnahmen aufklären.

3. Wenn unter Smogperioden die Belastung steigt, sollte man erwägen, die Öffentlichkeit, besonders die anfälligen Personen, genauer über Möglichkeiten der Expositions- und Dosissenkung zu informieren.

4. Kurzfristige Maßnahmen zur Senkung der Emissionen, die eine ununterbrochene Einbeziehung einer größeren Anzahl Menschen erfordert oder die Handlungsfreiheit vieler Bürger einschränkt, sollten nur in schwerwiegenden Smogperioden ergriffen werden. In solchen Fällen sollten genügend Haushaltsmittel vorhanden sein, um die Effektivität einer solchen Strategie auch im Hinblick auf eine echte Senkung der Gesundheitsrisiken zu evaluieren.

5. Die Meßstationen zur Beurteilung der Smogsituation sollten so verteilt sein, daß die Meßergebnisse die Exposition der örtlichen Bevölkerung widerspiegeln. Die Meßresultate sollten laufend abgefragt werden können, wobei der Zeitraum für die Durchschnittsbildung höchstens drei Stunden betragen darf, damit man für die nächsten 24 oder 48 Stunden extrapolieren kann.

6. Zur Erhöhung der Zuverlässigkeit von Smogprognosen sollte ein einfaches Modell verwendet werden, das einige charakteristische meteorologische Faktoren zu gemessenen Schadstoffkonzentrationen in Beziehung setzt.

7. Mitarbeiter des Gesundheitswesens und Angehörige anderer einschlägiger Berufe sollten so weit geschult sein, daß sie die Bürger in bezug auf die im Anhang beschriebenen gesundheitlichen Effekte beraten können.

Anhang I

Nachweislich mit Winter- und Sommersmog verbundene Gesundheitseffekte

Tabelle I: Über 24 Stunden gemittelte Konzentrationen von SO_2 und Schwebstoffe enthaltenden Schadstoffmischungen, bei deren Überschreiten aufgrund epidemiologischer Beobachtungen akute spezifische Auswirkungen auf die Gesundheit erwartet werden

SO_2	Schwebstoffe ($\mu g/m^3$)	Gesundheitliche Effekte	Klassifizierung der Gesundheitseffekte
200	200 (gravimetrisch)	Geringe, vorübergehende stufenweise eingeschränkte Lungenfunktion (FVC, FEV_1) bei Kindern und Erwachsenen, Dauer 2 – 4 Wochen. Größenordnung der Effekte 2 – 4% des Gruppendurchschnitts.	mittelmäßige
250	250 (dunkler Rauch)	Zunahme der Morbidität der Atemwege bei anfälligen Erwachsenen (chronische Bronchitis) und eventuell bei Kindern	mittelmäßige
500	500 (dunkler Rauch)	Zunahme der Sterblichkeit bei älteren und chronisch kranken Personen.	schwere

Anmerkung: Man nimmt an, daß die Schadstoffeinwirkung auf die menschliche Gesundheit bei Schwefeldioxid und Schwebstoffen ab 400 $\mu g/m^3$ bereits bedenklich wird.

Tabelle 2: Zu erwartende akute Gesundheitseffekte bei photochemischem Smog mit Angabe maximal einstündiger Ozonmittelwerte – gemäß toxikologischer klinischer und epidemiologischer Studien über Kinder und nichtrauchende Jugendliche (chronische Effekte nicht berücksichtigt)

Ozonkonzentration μ/m³	Reizung/ Entzündung von Augen, Nase und Hals (bei allen)	Durchschnittliche Reduzierung des FEV_1 bei aktiven Personen (im Freien)		Verbot von Aufenthalt und Tätigkeiten im Freien	Atemwegsreaktionen, Entzündungen, Clearance-Reaktionen, Hyperaktivität (bei im Freien aktiven Personen)	Symptome der Atemwege (hauptsächlich Erwachsene)	Klassifizierung der Gesundheitseffekte
		Gesamte Bevölkerung	die anfälligsten 10% der Bevölkerung				
< 100	kein Effekt	keine	keine	keine	keine	keine	
200	bei wenigen, empfindlichen Personen	5%	10%	keine	leichte	Spannung in der Brust, Husten	leichte
300	< 30% der Bevölkerung	15%	< 30%	einige Personen	mittelmäßige	Verstärkte Symptome	mittelmäßige
400	> 50% der Bevölkerung	25%	> 50%	einige Personen	schwere	Zusätzliche Verstärkung der Symptome	schwere

Резюме

Совещание рабочей группы было созвано для оценки различных видов риска для здоровья, ассоциируемых с эпизодическим воздействием повышенных концентраций загрязнителей воздуха зимой и летом ("зимний" и "летний" смог). Совещание было организовано при сотрудничестве и поддержке Министерства по жилищному вопросу, планированию физической среды и охране окружающей среды Нидерландов. В нем приняли участие 15 временных консультантов и 3 наблюдателя из восьми стран.

Участники рассмотрели последствия для здоровья человека влияния, которое точно установлено или которое ассоциируется с эпизодическим воздействием смога зимнего и летнего типа. В данном отчете подразумевается, что смог зимнего типа означает лишь загрязнение атмосферы в результате сжигания ископаемых видов топлива, содержащего серу. Обычно проводятся замеры концентрации двуокиси серы и взвешенных частиц в качестве индикаторных загрязнителей в данной смеси, хотя вредные последствия для здоровья человека могут в первую очередь быть и результатом воздействия других компонентов, например серной кислоты. Смог летнего типа связан прежде всего с фотохимическим загрязнением, являющимся результатом атмосферных реакций углеводородов и окисей азота, происходящих под воздействием интенсивного солнечного света. В этой смеси наиболее биологически активным загрязнителем считается озон.

Однако возможны значительные отклонения в составе смога как зимнего, так и летнего типа в различных местностях, и поэтому выводы, касающиеся воздействия в одном месте, не обязательно применимы к другой местности. Например, при сравнении мест, находящихся на различной широте с различной интенсивностью ультрафиолетового солнечного облучения и различными выбросами из источников загрязнений, токсичность фотохимического смога при тех же концентрациях озона может существенно отличаться. Кроме того, необходимо срочно уточнить характеристики воздействия загрязненного воздуха и связанные с интенсивным движением мототранспорта в центрах крупных городов последствия для здоровья воздействия выхлопных газов.

Некоторая информация, касающаяся последствий для здоровья, которые ассоциируются с воздействием эпизодически возникающего смога зимнего и летнего типа, приводится в Приложении 1. При повышенных уровнях воздействия эти последствия могут быть охарактеризованы как "слабые", "средние" или "сильные", понимая под этим, что при увеличении тяжести их воздействия более настоятельной становится принятие определенных мер.

Пиковые воздействия смога зимнего или летнего типа могут быть снижены в тех случаях, когда в ответ на тревожные сообщения по поводу загрязнения атмосферного воздуха принимаются меры к снижению выбросов, связанных с хозяйственной деятельностью человека и/или путем снижения воздействия в результате ограничения личных передвижений. Объявление тревоги по поводу загрязнения определяется как формальная акция соответствующего органа, вызываемая прогнозируемым превышением концентрации индикаторного загрязнителя по отношению к предельно допустимой концентрации, что окажет вредное воздействие на здоровье людей. Такого рода тревоги на практике могут случаться несколько раз в сезоне или в году, и они не приведут к каком-либо существенному снижению кумулятивного или долгосрочного среднего воздействия.

Официальным ответом на прогноз о превышении предельно допустимых концентраций в данном месте или при данной юрисдикции является возникновение ответственности у соответствующего органа. Когда ожидается воздействие слабого уровня, то единственной необходимой мерой представляется объявление о том, что предельно допустимые концентрации могут быть превышены, и разъяснение, что это означает для здоровья населения. Когда предполагается средний уровень воздействия, то может быть рассмотрен вопрос об определенных консультациях населения по поводу мер снижения такого воздействия или уменьшения его дозы на лиц, чувствительных к такому воздействию. Когда предполагаются тяжелые воздействия для здоровья, то могут быть рекомендованы дополнительные меры, принимаемые на добровольной основе. Такие чрезвычайные краткосрочные меры, как закрытие школ или наложение ограничений на движение транспорта, должны рассматриваться в тех случаях, когда есть вероятность превышения уровней, вызывающих тяжелые последствия для здоровья человека (Приложение 1, табл. 1 и 2).

Приведенные в таблицах уровни не представляют собой пороговые значения воздействия, но указывают на достаточно высокую степень загрязнения атмосферного воздуха, чтобы вызвать явления, которые могут быть обнаружены при соответствующих обследованиях. Более высокие уровни воздействия вызовут явления повышенной тяжести у большего числа людей, подвергшихся такому воздействию; однако невозможно указать конкретные числовые значения такого увеличения, основываясь на тех ограниченных данных, которые накоплены к настоящему времени. Уровень загрязнения, более низкий, чем самый нижний уровень, показанный в таблицах, не может считаться не оказывающим никакого влияния, однако предполагается, что он не вызовет такие явления, которые будут представлять собой крупную угрозу для здоровья людей.

В целом лица с легочными болезнями или недостатками кровообращения более тяжело, по сравнению с другими, переносят даже относительно небольшие по масштабам воздействия эпизодического смога зимнего типа. Что касается смога летнего типа, то четко не установлено, какие лица подвергаются особому риску, хотя хорошо известно, что некоторые люди более чувствительны к воздействию такого смога - и особенно озона - чем другие.

Группа в целом отметила, что факт созыва совещания по оценке риска для здоровья эпизодических явлений смога свидетельствует о том, что меры по снижению загрязнения не дали желаемого эффекта, и что результаты совещания никоим образом не должны отвлечь внимание ответственных органов от необходимости усилить действие по снижению базовых уровней загрязнения. При сниженных базовых выбросах пиковые концентрации, являющиеся результатом увеличения потребностей в энергии и ухудшения погодных условий, также снизятся. При таком подходе снижение пиковых концентраций даст возможность избежать наложения ограничений неотложного характера на хозяйственную и личную деятельность людей и будет иметь своим результатом дополнительное преимущество, связанное со снижением долгосрочного кумулятивного воздействия смоговых загрязнителей. В этой связи необходимо упомянуть об отчете, написанном Европейским региональным бюро ВОЗ для Европейской экономической комиссии ООН в Женеве, "Impact on human health of air pollution in Europe" (Влияние на здоровье человека загрязнения атмосферного воздуха в Европе)[a].

Опыт, накопленный в последнее время в нескольких европейских странах, показывает, что эпизоды сильного загрязнения атмосферного воздуха могут привлекать повышенное внимание средств массовой

---

[a] Данный отчет можно получить в Отделе по борьбе с опасными для здоровья факторами окружающей среды Европейского регионального бюро ВОЗ по адресу: Scherfigsvej 8, DK-2100 Copenhagen Ø.

информации и, следовательно, вызывать большую озабоченность населения, которая в целом не оправдывается тяжестью предполагаемого воздействия на здоровье человека. Однако психологический эффект и его воздействие на общее состояние людей, вероятно, сильно недооцениваются. В частности, те консультации, которые даются людям, подвергающимся значительному воздействию в процессе их обычной трудовой деятельности (например рекомендации избегать физической активности вне помещений и/или не покидать помещения), могут ассоциироваться, по крайней мере частью населения, с возникновением чрезвычайной ситуации. Крайне важное значение для поддержания необходимого доверия населения и сотрудничества с его стороны имеет эффективная информация об уровне риска для здоровья в связи с воздействием различных уровней загрязнения воздуха и об уровне загрязнения воздуха в любой момент времени. Это указывает на важное значение тщательной разработки программы образовательного характера для широких слоев населения.

ВЫВОДЫ

1. Несмотря на возможность выбора определенных "индексных" загрязнителей, характеризующих два основных рассматриваемых типа смога - например двуокиси серы и взвешенных частиц в случае со смогом зимнего типа и озона в случае смога летнего типа - могут наблюдаться значительные вариации в составе смесей загрязнителей в зависимости от местоположения, поэтому выводы, касающиеся их воздействия в одном месте, не обязательно применимы для другой местности.

2. Хотя контроль за выбросами и переход на другие виды топлива позволили ликвидировать наиболее опасные проблемы здравоохранения, действие которых обычно проявлялось при эпизодическом смоге зимнего типа, все еще в отдельных местностях случаются эпизоды повышенного загрязнения такого типа в результате неадекватного контроля за выбросами в соединении с метеорологическими и топографическими факто-

рами, что приводит к тяжелым последствиям для здоровья отдельных людей.

3. Эпизодические возникновения смога летнего типа за последние 20 лет стали обычным явлением во многих районах Европы, а их интенсивность и частота скорее увеличиваются, чем снижаются. В этой связи, основываясь на опыте других районов земного шара, следует ожидать тяжелых последствий для здоровья населения.

4. Пиковые воздействия смога зимнего или летнего типа могут быть снижены, если в ответ на тревожные сообщения о загрязнении атмосферного воздуха предпринимаются действия, результатом которых становится снижение выбросов, связанных с хозяйственной деятельностью, а также если принимать меры по уменьшению вредного воздействия смога путем ограничения передвижения людей. Однако наиболее предпочтительной альтернативой созданию систем объявления тревоги по поводу загрязнения представляется снижение пикового воздействия путем уменьшения базового воздействия, используя такие подходы, как переход на другие виды топлива, изменение соответствующих процессов и/или жесткий контроль за выбросами, которые снижают базовые объемы выбросов на достаточно больших территориях.

5. Системы предупреждения в отношении смога зимнего типа могут основываться на учете комбинаций содержания двуокиси серы и взвешенных частиц. В некоторых случаях может оказаться целесообразным использование превышений установленного уровня одного из этих веществ в качестве показателя, однако в других случаях необходимо использовать различные комбинации в зависимости от местных обстоятельств и соответствующей оценки риска для здоровья.

6. Учитывая различные уровни, используемые для отличия разных воздействий на здоровье эпизодических смогов зимнего и летнего типа, считается необходимым установить степень точности порядка 15% при определении фактического уровня индексных загрязни-

телей. В принципе число станций, необходимых для удовлетворения требований к точности измерений, будет зависеть от объемного градиента распространения загрязнения над рассматриваемым районом.

РЕКОМЕНДАЦИИ

1. В отдельных районах в зависимости от эпизодов смога зимнего или летнего типа мониторинг должен распространяться на более широкий круг загрязнителей, а не на один базисный "индексный" загрязнитель, с тем чтобы дать характеристику всей смеси загрязнителей; там, где это возможно, эпидемиологические исследования должны проводиться на месте, чтобы обеспечить информацией для оценки риска для здоровья населения.

2. Соответствующие органы должны предпринять действия по информированию широких слоев населения о потенциальных серьезных последствиях для здоровья эпизодического возникновения смога зимнего и летнего типа, в зависимости от масштабов этой проблемы, и должны сообщить им, какие меры могут быть предприняты отдельными лицами для их защиты от неблагоприятных последствий, связанных с такими эпизодами.

3. При увеличении уровней воздействия во время таких эпизодических смогов необходимо рассматривать вопрос о конкретных рекомендациях относительно возможности сокращения времени воздействия такого смога или снижения дозы воздействия на чувствительных к смогу лиц.

4. В тяжелых ситуациях при густом смоге необходимо опираться лишь на стратегии краткосрочного снижения числа источников загрязнения, что требует постоянного сотрудничества большого числа людей или налагает ограничения на свободу выбора многих людей. Для того чтобы такая стратегия работала, необходимо иметь достаточные бюджетные средства для оценки эффективности стратегии в форме реального сокращения масштабов риска для здоровья.

5. Измерительные станции, используемые для оценки смогов эпизодического характера, должны располагаться таким образом, чтобы результаты измерений давали представление о масштабах воздействия смога на население рассматриваемого района. Должна быть налажена постоянная связь для получения сообщений о результатах измерений при средних интервалах самое большее в три часа, чтобы иметь возможность составить прогноз на следующие 24 или 48 часов.

6. Для улучшения качества прогнозов о наступлении смога необходимо пользоваться простой моделью, в рамках которой некоторые характерные метеорологические факторы объединяются с замеренными уровнями загрязнения.

7. Работники системы здравоохранения и сотрудники других отраслей, имеющие отношение к этим вопросам, должны обеспечиваться достаточно полной информацией, которая позволит им дать квалифицированную консультацию любому человеку относительно того, какие последствия для здоровья связаны с величинами, приведенными в Приложении 1.

Приложение 1

КРАТКИЙ ПЕРЕЧЕНЬ ПОСЛЕДСТВИЙ ДЛЯ ЗДОРОВЬЯ, АССОЦИИРУЕМЫХ СО СМОГОМ ЗИМНЕГО И ЛЕТНЕГО ТИПА

Табл. 1. Уровни среднесуточных концентраций смесей загрязнителей атмосферного воздуха, содержащих SO_2 и взвешенные частицы, превышение которых предполагает наступление конкретных тяжелых последствий для здоровья человека, основываясь на результатах наблюдений, полученных в результате эпидемиологических исследований

SO_2	Частицы (мкг/м^3)	Последствия для здоровья	Общая классификация степени воздействия
200	200 (гравиметрический показатель)	Небольшие кратковременные ухудшения функционирования легких (FVC, FEV_1) у детей и взрослых, которые могут продолжаться в течение 2-4 недель. Масштабы нарушений составляют порядка 2-4% от среднегрупповых значений.	Средняя
250	250 (сажа)	Увеличение числа легочных заболеваний среди восприимчивых к ним взрослых (хронический бронхит) и, возможно, детей.	Средняя
500	500 (сажа)	Увеличение смертности среди пожилых и хронически больных людей	Тяжелая

Примечание: Тяжелыми считаются воздействия на здоровье человека при уровнях 400 мкг/м^3 как для двуокиси серы, так и для взвешенных частиц.

Табл.2 Ожидаемые тяжелые воздействия фотохимического смога в дни, характеризуемые максимально одночасовыми средними концентрациями озона, определяемые для детей и некурящих молодых людей на основе наблюдений при проведении токсикологических клинических и эпидемиологических исследований (хронические эффекты не рассматриваются)

Уровень озона $мкг/м^3$	Раздраже- ние глаз, носа и глотки (у каждого)	Среднее снижение FEV_1 у актив- но двигающихся людей (вне помещений)		Наложение огранич.на время пре- бывания вне поме- щений и активную деятель- ность	Воздействие на органы ды- хания, возни- кновение вос- палительных процессов и клиренса, гип- перреактивно- сть у активно двиг. людей вне помещений	Респиратор- ные симптомы (в первую очередь у взрослых)	Общая кла сификаци воздействи
		Все нас- лен- ие	Наиболее чувствит. часть - 10% населения				
<100	Никакого воздейст.	Нет	Нет	Нет	Нет	Нет	
200	У неболь- шого числа чувствит. людей	5%	10%	Нет	Слабая степень	Некоторая стесненность в груди, кашель	Легкая степень
300	<30% на- селения	15%	<30%	Отдельные люди	Средняя степень	Усиленные симптомы	Средняя степень
400	>50% на- селения	25%	>50%	Большое число людей	Тяжелая степень	Дальнейшее усиление симптомов	Тяжелая степень